MW00888467

Bombo's Big Question

By Kip Will, Illustrated by Ainsley Seago

Copyright © 2018 Kipling Will
All rights reserved.
ISBN-10: 1986476413
ISBN-13: 978-1986476416

Acknowledgments
The original concept for this story by Sara Marie Sweer.

Thanks to Jonathan Tweet, Anne H. Weaver, and Roberta Brett for very useful comments and suggestions that significantly improved the story.

 This project is supported by funding from the US National Science Foundation, DEB1556957 DEB1556813, DEB1556931, DEB1556898

In a woodland along a beautiful stream lived a bombardier beetle named Bombo.

Bombo was young and full of wonder and curiosity. She was very eager to explore her world.

To learn more about beetles see page 25.

Just as she set out to explore, she was attacked!

Alas, poor Bombo!

A hungry frog wanted to make Bombo his snack. But...

...without thinking, Bombo fought back!

Learn more about this frog on page 26.

With a loud *"pop!"* a stream of boiling hot fluid hit the frog square in the face!

The hot, foul-tasting blast startled and distracted the frog and Bombo was able to escape!

3

Once she was a safe distance away, Bombo paused to wonder about this mysterious weapon that had saved her from being eaten by the frog.

It was then that she realized that she was the weapon. She had the ability to defend herself with a chemical blast from her own body!

But how? Where did this ability come from?

Bombo was once again filled with wonder and curiosity.

The next night, Bombo set out to find answers to these questions about her stunning defensive ability.

First she came upon a dandy *Chlaenius* beetle, glimmering metallic in the moonlight, and asked him,

"Hello *Chlaenius*, can you tell me how it came to be that I can spray a hot chemical blast at my enemies?"

What kind of beetle is *Chlaenius* and how do I say that name? Find out on pages 25-26.

5

Chlaenius groomed his antennae thoughtfully, and said,

"I don't know, but I too can spray chemicals, as do many of our kin.

"Though clearly I smell the best, there's so much variation. Some of our cousins have chemicals that make ants go crazy, others can turn toads. We're all a little bit alike with our chemicals, but also a little different."

With that, *Chlaenius* scuttled under a rock.

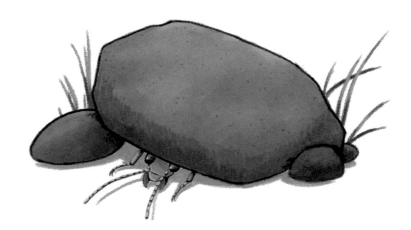

6

Bombo kept walking, and along the way she met a firefly with her bum all aglow. Bombo asked,

"Excuse me Firefly, you seem very bright. Can you tell me how it came to be that I can spray a hot chemical blast at my enemies?"

But the firefly dismissed Bombo's question, and seemed annoyed by her curiosity.

"Shh," said female *Photuris* Firefly. "Can't you see by my flashing light that I'm hunting *other* fireflies to eat?

Find out what this firefly really eats on page 26.

"You're interrupting my dinner. Anyway, I'm sure you just inherited the ability to spray chemicals from your ancestors, just like I inherited my light".

The next evening Bombo saw a fuzzy Marsh Rat with a long, graceful tail that helped it to balance as it moved.

She asked the rat, "Hello Rat, you are a clever mammal. Can you tell me how it came to be that I can spray a hot chemical blast at my enemies?"

What is a Marsh Rat? Find out on page 26.

9

The Marsh Rat squeaked proudly: "Sometimes things are just the way they are.

"You need not be so curious, nor ask so many questions. You only need to believe you are the way you were made to be."

But Bombo was a smart, curious beetle, and she did not think that sounded like a good answer at all.

On the third night, Bombo was feeling a little discouraged.

Would she ever get an answer to her big question, or would she have to wonder forever about how she got her amazing ability?

She had some clues from *Chlaenius* beetle and *Photuris* firefly, but something was missing.

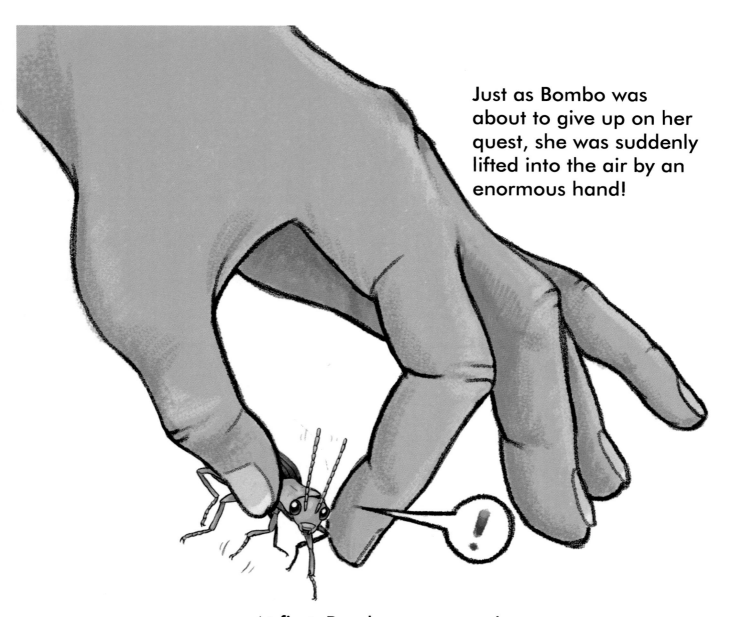

Just as Bombo was about to give up on her quest, she was suddenly lifted into the air by an enormous hand!

At first, Bombo was scared and was just about to blast her defensive chemical spray...

...but when it was clear she wasn't going to be eaten, Bombo took a deep breath and asked the giant her big question:

"I-I have to know! Can you tell me how it came to be that I can spray a hot chemical blast at my enemies so I can escape from them?

"I learned from *Chlaenius* that there is variation among different beetles' chemistry, and that some chemicals work better on different kinds of enemies, and *Photuris* Firefly suggested that I inherited this power from my ancestors.

"But how do these things all fit together?"

14

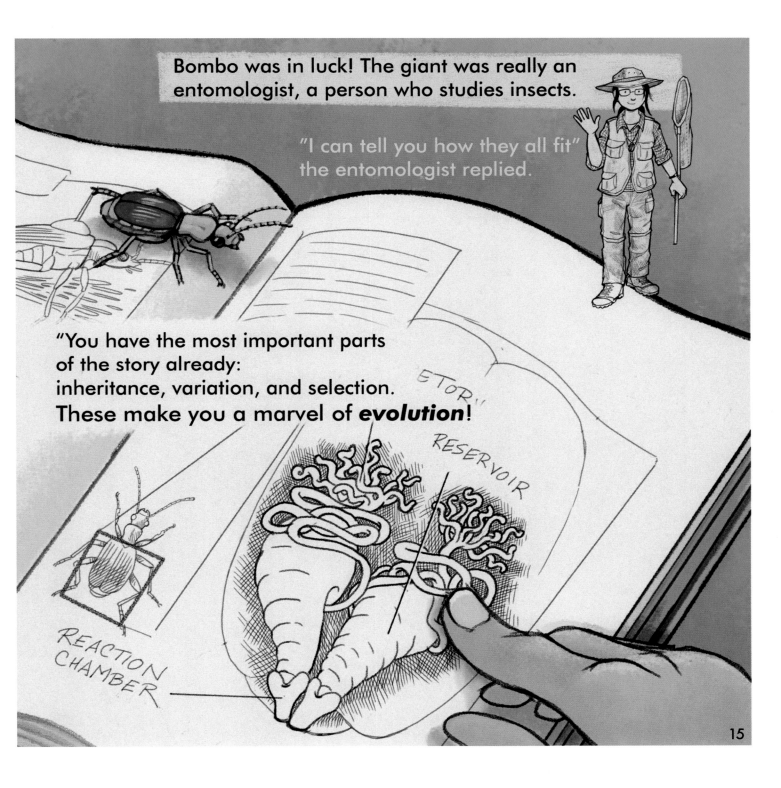

Bombo was in luck! The giant was really an entomologist, a person who studies insects.

"I can tell you how they all fit" the entomologist replied.

"You have the most important parts of the story already: inheritance, variation, and selection. These make you a marvel of *evolution*!

RESERVOIR

REACTION CHAMBER

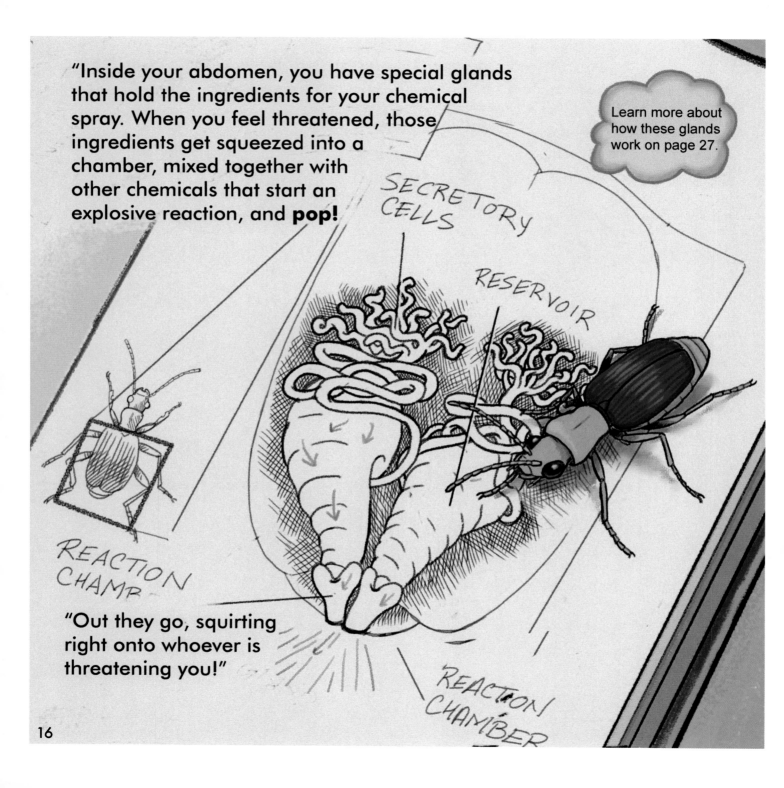

"Inside your abdomen, you have special glands that hold the ingredients for your chemical spray. When you feel threatened, those ingredients get squeezed into a chamber, mixed together with other chemicals that start an explosive reaction, and **pop!**

Learn more about how these glands work on page 27.

SECRETORY CELLS

RESERVOIR

REACTION CHAMBR

"Out they go, squirting right onto whoever is threatening you!"

REACTION CHAMBER

"Wow!" Bombo exclaimed.
"That's amazing!

"How did evolution give me this superpower?"

RESERVOIR

REACTION
CHAMBER

GLAND

"Well," the entomologiest replied, "you can thank natural selection and inheritance from your ancestors for your superpower. Just like you, they needed to defend themselves against frogs, ants, and other pesky predators.

"Over many generations, the genes for different kinds of chemicals were passed down through different branches of the beetle family tree.

18

"Now you have thousands of cousins out there who make similar chemicals, for the same reason— some spray hot blasts like yours, some produce cooler ones, but all help to repel predators!"

Learn more about family trees on page 27.

GROUND BEETLE FAMILY TREE

Bombo was amazed and filled with wonder.

She had one more question for the entomologist: "But how did you learn about *my* ancestors if they're, well, my ancestors? Are you millions of years old?"

The entomologist smiled kindly, and replied, "I may be older than you, but not that old! However, the Earth is a great deal older than we are, and it keeps records of much of what has happened.

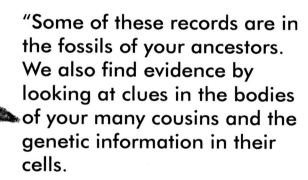

"Some of these records are in the fossils of your ancestors. We also find evidence by looking at clues in the bodies of your many cousins and the genetic information in their cells.

Learn more about cells and genetic information on page 28.

POWER OF methacrylic acid!

POWER OF dimethyltrisulfide!

POWER OF salicylaldehyde!

POWER OF formic acid!

POWER OF m-cresol!

"We can see how you and your beetle cousins share similarities, including defensive chemical superpowers!"

Chemistry rocks! See more about these beetles and their compounds on page 28.

Bombo thanked the entomologist as she waved goodbye and was very happy she had met her.

She had all these wonderful thoughts about her place in life on Earth and knowledge that satisfied some of her curiosity.

Most of all, she had the powerful explanation that evolution offered for her fantastic ability.

What is Bombo still curious about? Learn about our current research on page 29.

Bombo found a good answer to her big question.

Background and additional information

How do you say that?

Pronunciation of names: Here are all the scientific names used in the story and in the additional information below. Scientific animal names are made of Latinized words and can be unfamiliar to many people. Pronunciation might seem challenging, but because names are very precise and unique (one per species) they are very important. Here is a guide if you want to pronounce the names used in this book the way scientists do.

Anthia. (ăn-thī'-ä) Saber-toothed ground beetles.
Brachinus. (bră-kī'-nŭs) Common bombardier beetles.
Calosoma. (kăl-ō-sōm-ä) Caterpillar hunter beetles.
Carabidae. (kă-ră'-bĭ-dē) Ground beetles.
Chlaenius. (klē'-nē-ŭs) Vivid metallic ground beetles.
Lithobates catesbeianus. (lĭthō-bā'-tēs că-tĕs-bē-ĭ'-ăn-ŭs) Common Bullfrog.
Oryzomys palustris. (ō-rē-zō'-mīz pä'-lŭs-trĭs) Marsh Rice Rat.
Panagaeus. (pă-nă'-jē-ŭs) Rood mark beetles.
Photinus. (fō-tī'-nŭs) Rover fireflies.
Photuris. (fō-tûr'-ĭs) Common fireflies.
Pheropsophus. (fĕ-rŏp'-sō-fŭs) Noisy bombardier beetles.
Pterostichus. (tĕ-rŏ'-stĭ-kŭs) Woodland ground beetles.
Psydrus piceus. (sĭ-drŭs pī'-cē-ŭs) The Sulfur Beetle.

Beetles: Bombo is a beetle. So are *Chlaenius* and *Photuris* firefly. Beetles (the order Coleoptera) are the largest natural group of insects. All beetles, including bombardier beetles, have a life cycle like butterflies that includes an egg, several juvenile stages (larvae), a pupa, and an adult stage. In our story, Bombo is an adult beetle that has just emerged from her pupal stage. Bombo is representative of a particular family of beetles (Carabidae) that includes bombardiers (the genera *Brachinus*, *Pheropsophus*, and others) and non-bombardier beetles, but all of these beetles use chemical defenses of one kind or another. About 500 species of bombardier beetles like Bombo are found throughout much of the world. There is a good chance that they are living along a stream near you.

Chlaenius beetles: Like Bombo, *Chlaenius* is a kind of beetle that is found in many parts of the world. They are usually brightly colored and are fierce predators. Some even turn the tables on frogs and will eat them. They also have very strong chemicals. Some species of *Chlaenius* use the same chemical as bombardiers (quinones) but don't have the ability to make hot spray. If you look around ponds and wet places you might see *Chlaenius* out hunting.

Fireflies: The *Photuris* firefly in the story says she is hunting other fireflies. In fact these beetles hunt and eat males of a different genus, *Photinus* fireflies! In this case the female *Photuris* firefly uses a pattern of flashing that looks like the signal from a friendly female *Photinus*. When a male *Photinus* comes over to say hello she grabs him and gobbles him up. *Photuris* and *Photinus* also have chemical defenses, but rather than spray them, the chemicals remain inside their bodies. The chemicals don't stop *Photuris* firefly from eating other fireflies, however.

Hungry Frog: The Common Bullfrog (*Lithobates catesbeianus*) is a native species in North America that has been introduced into South America, Europe, and Asia. They live in and around wetlands and streams just like Bombo and other bombardier beetles. These frogs can grow to be very large (up to 0.5 kg) and have voracious appetites. These frogs live 7-9 years and eat many insects during their life, especially if the insects aren't chemically defended like Bombo.

Mouse: The rat in our story is the Marsh Rice Rat (*Oryzomys palustris*). They are found in the wetlands of the southeastern United States, where they live with beetles like Bombo. They are active at night and are very nimble swimmers and climbers. Their swimming and climbing abilities are partly due to their long tails.

How bombardiers blast: Bombardier beetles like Bombo and all of her close relatives (the family Carabidae) have a pair of glands inside their abdomen. Each gland consists of a cluster of special cells that secrete two kinds of quinones and hydrogen peroxide. Those compounds are stored together in a reservoir and do not react with each other. There is no chance of the beetles suddenly blowing themselves up. The reservoir has a strong muscle on the outside and a tight valve that is kept closed until they release the spray. In the story, when Bombo was attacked she squeezed those muscles, opened the valve, and pushed some of the fluid from the reservoirs into a structure called the reaction chamber. In the reaction chamber special proteins called enzymes, which were produced by accessory glands on the outside of the chambers, allow the chemicals in the fluid to transformed other compounds, producing heat and oxygen gas. The reaction is super fast and most of the chemicals are sprayed on the target so Bombo isn't hurt by her own blast.

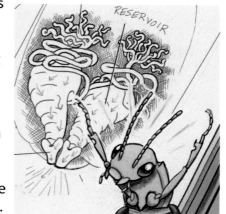

Family trees and evolving characters: All of life on Earth is related and shares a common ancestor. Because of this fact we can roughly depict the ancestry of any organism as a tree (a phylogeny) similar to

how we think of the genealogy of our own families. We base the relationships in our phylogeny on evidence from shared characteristics of the anatomy and genetics of the organisms. Bombo shares many characteristics, including the glands that produce defensive chemicals, with her relatives. These other species that are close relatives share the glands, but may produce different chemicals. Not all beetles live in the same place and over time they face different predators. These different situations and the natural variation found in any population are fundamental to how evolution works.

Genes and cells: Our bodies and the bodies of beetles like Bombo are made up of cells. Tissues are made up of groups of cells that have a particular function and organs, like the defensive gland system in beetles, are a collection of tissues that have one or a few tasks. All of their functions are controlled by genes that are made of DNA. Genes in cells also make or control the production of compounds like those in the bombardier glands. The DNA itself also changes over time in a lineage. We can look at changes, similarities, and differences between species, and use this as evidence for their relationships.

Chemicals and chemical diversity in beetles: Almost everything is made of chemicals! Chemists do research to understand chemicals and how they interact. This is a very important area of study. Bombo and her relatives have particular chemicals that are stored in strong concentrations and are unpleasant for any predator that attacks. It would be like someone spraying strong vinegar on a delicious piece of candy you were just about to eat. We list just a few of the hundreds of chemical compounds beetles produce. They

aren't delivered with a hot explosion the way Bombo does it but are sprayed or oozed. Even though they are not hot, they are bitter and irritating to a predator.

We show in the story (page 20) some examples of beetles and their chemical compounds . These beetles are fairly common and you can find more information about them in other books and online sources.

Methacrylic acid makes the Woodland Ground Beetle (*Pterostichus* species) smell like stinky cheese, and old leather. Formic acid is a sharp smell and the Saber-toothed Ground Beetle (*Anthia* species) can squirt it in your eye. Ouch! Dimethyltrisulphide— pee-ew! Who farted? This smelly stuff is made by the well-named Sulphur Beetle (*Psydrus piceus*) and no one wants to eat them. Salicylaldehyde is an aromatic compound with a name that is a mouthful to say but it tastes so bad that you don't want a mouthful of Caterpillar Hunter Beetles (*Calosoma* species), a beetle that makes this chemical. M-cresol is a rubbery-tar smelling compound that is nothing like the smell of food for most animals. It's produced by the brilliantly colored Rood Mark Beetle (*Panagaeus* species).

Bombardier chemical biology research and the scientific approach to life: Bombo is a smart beetle and she found that evolution by natural selection was a really good answer to her big question, but she remains curious about the natural world and how it works. There is always more to learn. Only by asking questions, testing what we think we know, and requiring evidence for our conclusions can we truly answer important questions about the world around us.

While we know a lot about bombardier beetles, there are many details to discover and questions still to answer. With new techniques and building on the solid evidence from the past, the current **Carabid-Q team** is exploring the detailed connections between the bombardier beetles' evolution, chemical biosynthesis, and the underlying genetics. We, like Bombo, are always curious and asking questions about the natural world.

Carabid

We plan to write the next chapter in Bombo's story and we want you, young scientists, to stay full of wonder and curiosity so that you can find really good answers to your big questions.

Wendy Moore, a professor at the University of Arizona, is an evolutionary biologist and insect systematist with an inordinate fondness for bombardier beetles. *"If you think bombardier beetles are cool, myrmecophilous bombardier beetles will blow your mind!"*

Kip Will, a professor at UC Berkeley, is an insect systematist and carabid beetle researcher. He is devoted to understanding the natural history of carabid beetles through lab experiments and field expeditions. *"Wherever carabid beetles roam, so do I."*

Reilly McManus, graduate student and lab manager at the University of Arizona, researches the ecological and evolutionary relationships between insects and their symbiotic microbial communities. *"In this microbial world, none of us have to make it alone. We get by with a little help from our friends."*

Ramu Errabelli, a postdoctoral researcher at Stevens Institute of Technology, Hoboken, is a synthetic and analytical chemist. *"Synthetic chemistry is the best kind of bartending."*

Adam Rork, a Ph.D. student at the The Pennsylvania State University, is interested in understanding the evolution of chemical defensive strategies in carabid beetles, particularly among bombardier beetles and formic acid-producing species. *"Biochemistry is the defensive toolbox of evolution."*

Athula Attygalle, a faculty member at Stevens Institute of Technology, Hoboken is a natural products scientist and insect chemistry researcher. He is devoted to understanding the role of chemicals arthropods use for communication and defense. *"Chemical warfare is not prohibited in the insect world."*

Tanya Renner, a professor at The Pennsylvania State University, is an evolutionary biologist with research interests in the chemical ecology of defense. She seeks to understand why and how enzymes evolve new functions, as well as how changes at the molecular level affect interactions between organisms. *"DNA is the macromolecule that links all life on Earth, from begonias to beetles."*

Aman Singh Gill, a postdoctoral researcher at UC Berkeley, is an evolutionary ecologist interested in insects, microbes and genomes in general. *"If there's no way to disprove an idea then it lives in the realm of faith, not fact."*

Zhaoyu Zheng, a Ph.D. student of Chemistry at Stevens Institute of Technology, works on mass spectrometric methods for structural identification of analytes. Her areas of interest include Laser desorption/ionization of inorganic materials, fragmentation patterns that reveal periodic trends and ion-mobility mass spectrometry. *"I can write a paper on a pencil."*

Sihang Xu, a PhD student at Stevens Institute of T echnology. His research focuses on chemical secretions from carabid beetles and ants, fragmentation pattern of small molecules in gas phase, and proton transfer in protomers by ion-mobility spectrometry mass-spectrometry technique. *"Insects are small great puzzlists."*

The Shoulders of Giants

Since the 1960s scientists have been actively researching bombardier beetles to build an evidence-based, scientifically tested explanation of their abilities. Much about these beetles and their evolution is now well understood. They represent one of the best cases in support of natural selection we have. The current state of understanding is a credit to many excellent scientists. Here are few of them.

Thomas Eisner (1929-2011) was a Professor of chemical ecology at Cornell University, a pioneer of insect chemical ecology and of deciphering the mysteries of bombardier beetles and many other insects. "*Insects are the most versatile chemists on Earth.*"

Jerrold Meinwald (1927-2018) was a Professor of Chemistry at Cornell University. He collaborated with Thomas Eisner for over a half century, showing how animals and plants use organic chemistry to communicate and defend themselves. Tom and Jerry enjoyed playing thousands of hours of chamber music together as well.

Hermann Schildknecht (1922-1996) was a Professor at the University of Heidelberg whose research into the chemical defense of bombardier beetles established the evidence for the chemical reactions as we know them now.

Terry Erwin, a Research Entomologist and Curator at the National Museum of Natural History, Smithsonian Institution, earned his doctorate with a monograph on bombardier beetles and since has devoted years on-site in the Neotropical rainforest studying carabid beetles of the forest canopy. "*Tropical forest canopies, the last biotic frontier.*"

Murray S. Blum (1929-2015) was a Professor of entomology at the University of Georgia, a pioneering researcher into the many surprising ways that insects use chemistry for everything from defense to communication, and the author of Chemical Defenses of Arthropods. "*Beetles are such good chemists that they ought to qualify for a Nobel Prize.*"

Barry Moore (1925-2015) was a chemist and expert on Australian carabid beetles. His work looking comparatively at defensive chemicals of beetles established our first classification of the chemicals they produce.

Konrad Dettner, a retired Professor at University of Bayreuth, Germany is a biologist and natural product chemist researcher. He is devoted to natural history and chemical ecology of insects and especially beetles. "*If beetles possess defensive glands or are toxic, I am always fascinated.*"

Daniel Aneshansley is a Professor in the Department of Biological and Environmental Engineering. He works on projects relating to the development of sensors and techniques to measure biological phenomena and the development of engineering models to study animal physiology.

Made in the USA
Las Vegas, NV
15 November 2023

80546159R00024